Newton and Gravity

THE BIG
iDEA

NEWTON

AN ANCHOR BOOK
PUBLISHED BY DOUBLEDAY
a division of Bantam Doubleday Dell Publishing Group, Inc.
1540 Broadway, New York, New York 10036

ANCHOR BOOKS, DOUBLEDAY, and the portrayal of an anchor are
trademarks of Doubleday, a division of Bantam Doubleday Dell
Publishing Group, Inc.

Newton and Gravity was originally published in the United Kingdom by
Arrow Books, a division of Random House U.K. Ltd. The Anchor Books
edition is published by arrangement with Arrow Books.

Library of Congress Cataloging-in-Publication Data

Strathern, Paul, 1940–
Newton and Gravity/Paul Strathern.
p.   cm.—(Big idea)
Includes bibliographical references.
1. Newton, Isaac, Sir, 1642–1727—Biography. 2. Gravity.
3. Physicists—Great Britain—Biography. I. Title. II. Series.
QC16.N7S77   1998
530′.092—dc21
[B]   97-52135
CIP

ISBN 0-385-49241-3
Copyright © 1997 by Paul Strathern
All Rights Reserved
Printed in the United States of America
First Anchor Books Edition: August 1998

1   3   5   7   9   10   8   6   4   2

Contents

Introduction

A GOOD CASE can be made for Newton being the finest mind humanity has yet produced. Shakespeare used language as no other, Napoleon used personality as no other—but no one has ever extended the limits of human understanding quite so drastically as Newton.

His work represents an evolutionary advance in our thinking—one giant leap for mankind. Long before we landed on the moon (or even considered doing so), Newton's mathematics paved the way for such a feat. Before Newton, the moon was part of the heavens, subject to unknown heavenly laws of its own. After him it became a satellite of earth, kept in orbit by the planet's gravitational pull. Humanity had its first glimpse of how the entire universe worked.

But the discovery of universal gravity was only the most major of Newton's many major discoveries. The concept of force, calculus, the nature of light, the theory of mechanics, the binomial series, Newton's method in numerical analysis—the list goes on and on. More units, and scientific and mathematical entities, are named after Newton than after any other scientist. The newton (the si, or internationally agreed unit of force), Newtonian fluid, Newton's formula (for lenses), Newton's rings (in optics), the Newton quotient (in differentiation) and many more—each the direct result of his work.

Yet all this was only possible because Newton arrived at the right moment in history. Just as Dante could only have written his *Divine Comedy* within the rigid all-embracing hierarchy of the Middle Ages, so Newton could only have made his discoveries after Copernicus and Galileo had freed the scientific mind from those same rigidities. As Newton himself confessed: "If I have seen a little farther than others it is because I have stood on the shoulders of giants."

The chains of medieval restraint had fallen away, and the door of human knowledge had opened upon a new world. In Newton's view, he had achieved little: "I seem to have been only like a boy playing on the seashore, and diverting myself in now and then finding a smoother pebble or a prettier shell than ordinary, whilst the great ocean of truth lay all undiscovered before me." The modesty exhibited here is of course dwarfed by the oceanic vision—which he alone was in a position to see. An implication that Newton may well have intended. He was not by nature modest.

So what kind of man was the possessor of the greatest intellect in history? In general, Newton's contemporaries came to regard him much as we in the twentieth century regarded

Einstein. A tame eccentric, member of a rare protected species, the absentminded genius of unquestionable moral stature: a distant but faintly lovable figure—vouchsafed enormous gravitas by the sheer weight of his achievements. In his time, Newton was the solitary scholar chosen by his peers as MP for Cambridge University, the revered president of the Royal Society reelected unopposed year after year, the Master of the Royal Mint feared and hated by the counterfeiters of the London underworld. As is so often the case, it was the common people who recognized the man for what he was. For beneath the austere public facade lay a deeply disturbed and vindictive personality, harboring his own illicit secrets.

Newton and Gravity

Life and Works

ISAAC NEWTON was born on Christmas Day 1642 in the manor house at the hamlet of Woolsthorpe in Lincolnshire. By coincidence his great scientific predecessor Galileo had died earlier in the same year.

Nowhere in the Newton family tree is there any sign of exceptional predecessors. His father, also Isaac Newton, was a prospering yeoman who could not even sign his name. "A wild, extravagant, weak man," according to his family, he died three months before the birth of his son. His mother was the daughter of an impecunious local gentleman, and was generally regarded as a hardworking, frugal woman.

Isaac was born prematurely, and was "so little they could fit him into a quart pot." He was not expected to survive the day of his birth. (As it turned out, he was to have exceptional health and live for eighty-four years.) Having never known his father, young Isaac was to "lose" his mother when he was just eighteen months old. In 1644 Hannah Newton married the sixty-three-year-old Barnabas Smith, a well-off local minister, and went to live in the village of North Witham. Young Isaac was left behind, to be looked after by his grandmother.

Newton never forgot this traumatic event, and its effect left an indelible imprint on his character. His adult life was to be marred by uncontrollable rages, paranoid vindictiveness, and occasional mental instability. He loved his mother, but she had abandoned him. He couldn't bring himself to hate her, but heaven help anyone who presented a legitimate target on which he could vent his repressed inner fury.

In fact, North Witham was only a couple of miles up the valley. Young Isaac could even see its church tower across the fields from the hill above his home. But in reality it remained a world away. His true Father was "in heaven," and his mother withdrawn to the limits of his childhood world. In adult life Newton was to devote himself to long and profound speculation about distant heavenly bodies and the nature of their attraction to one another. Not surprisingly, psychologists have seen more than unalloyed coincidence at work here.

According to a contemporary, Newton grew up a "sober, silent, thinking lad." But he was also subject to occasional outbursts of "tantrums." During one of these, Newton was later to remember "threatening my father and mother Smith to burn them and the house over them." So it seems that initially his mother wasn't always spared his wrath. (And pyromania, even in wishful form, hardly betokens normalcy.)

But Newton's mind wasn't the only thing on fire. In the year of his birth the behavior of Charles I and his belief in the divine right of kings finally drove the Parliamentarians to challenge his rule. The ensuing civil war raged all over England during the first six years of Newton's life, ending in victory for the Parliamentarians and the execution of Charles I in 1649. Throughout the civil war, sporadic fighting and house-burning took place in Lincolnshire. Newton and other local landowning families were inclined to support the king, but not to the extent of taking up arms.

The Parliamentary victory—the first successful revolution in Europe—saw the establishment of the Commonwealth, followed by post-revolutionary excesses such as have now become the norm. A repressive Puritanism was enforced. All dancing and displays of public merriment were banned, and even Christmas became a day of prayer rather than pudding eating. Yet here again the farming families of Lincolnshire were hardly affected. They had long lived austere, God-fearing lives, with the emphasis on Bible reading and the shocking-ness of sex. Young Isaac grew up in a habitu-ally Puritan household, and absorbed Puritan habits as a matter of course. He learned to consult the Bible to discover the wishes of God the Father, a habit he was to retain through-out his life.

But God the Father was not only God in heaven, He was also Father in heaven. In the ever-booming field of Newtonian psychologi-cal studies, most are agreed that Newton was driven by a strong unconscious need to know his Father. He knew from his faith that God the Father had made the universe, leaving cer-tain clues as to its ultimate nature and His ulti-mate intentions. Throughout his life Newton was driven to search obsessively for these clues—in the two appropriate fields. He was to devote just as much of his time to pursuing biblical and religious studies as he was to the pursuit of scientific truth. And to the end, he was convinced that his religious work would have the most lasting value. For once the facts appear to be as mad as the psychology.

When Newton was ten, the Reverend Barnabas Smith died and Isaac's mother returned home to Woolsthorpe a comparatively rich woman. Newton's prayers had been answered. There followed two years of curious bliss, tempered by the common-sense reality of his mother, and the additional presence of a half brother and two half sisters. But Isaac was the eldest, and Hannah seems to a certain extent to have relied upon him. Before Isaac had even reached puberty he had become "the man of the family" in his mother's eyes. The basic self-belief engendered by this precocious motherly recognition was never to desert him in his intellectual endeavors, even when the man himself was beset by maddening anxieties.

At the age of twelve, Newton went to the grammar school in Grantham, which was ten miles away. Here he took lodgings with Mr. Clark the apothecary, whose house was on the High Street beside the George Inn. At school his studies consisted almost entirely of Latin and ancient Greek. Mathematics was all but ignored in the education of the period, which remained for the most part medieval. The quiet, sensitive country boy was uninterested, and sank to the bottom of the class.

According to his own account, Newton remained intellectually dormant until the day when he was kicked in the stomach by the school bully. Newton challenged him to a fight in the churchyard. In the words of Newton's first biographer Conduitt, who recorded Newton's reminiscences: "Isaac was not so lusty as his antagonist [but] he had so much more spirit and resolution that he beat him till he declared he would fight no more." Newton had found a legitimate target on which to vent the dreadful anger that lay repressed within him. But once this was aroused it became uncontrollable, and then there was no holding him back. Trouncing his physically superior opponent was not enough. After his victory "Isaac pulled him along by the ears and thrust his face against the side of the church and rub his nose against the wall." But even such physical humiliation was not sufficient. Newton had to defeat his opponent in every way possible. He felt the need to better his opponent intellectually, began trying in class, and was soon demonstrating his intellectual superiority.

This is the way Newton remembered it, and there is no doubt that something very similar happened. Such anger-fueled vindictiveness was to recur at intervals throughout his life: This outbreak merely set the pattern.

Once Newton's intellectual faculties had been roused, there was no stopping him. Watching the teenage dullard emerge from his chrysalis and stretch the butterfly wings of genius must have been a wondrous sight for the townsfolk of Grantham. And, of course, they all remembered it. In hindsight. According to reminiscences collected after the death of the great Sir Isaac Newton, president of the Royal Society, Master of the Royal Mint etc., young Isaac displayed all the expected signs of supreme genius—baffling the locals with intricately constructed model windmills, handmade water clocks, exploding kites, a mouse-driven corn mill, a foldable paper lantern, his ability to tell the precise time from a shadow, and a notebook filled with the usual unintelligible diagrams. Fortunately this notebook is now in the Pierpont Morgan Library in New York, an inscription inside the front cover recording that it was originally bought by Newton for $2^{1}/_{2}$d (i.e., two and a half old pennies) in 1659. Its contents confirm the seemingly fanciful memories of the people of Grantham, with pages containing diagrams of Copernicus' solar system, details of how to make a sundial and construct a model windmill, and astrological predictions of eclipses. Two things are obvious. Newton's intellectual interests had expanded far beyond the limits of his school education. And his main interest was in science and how things worked.

All the evidence points to a precociously brilliant, largely self-taught amateur. Unusual, but not unique. There must have been a score or more of similar prodigies throughout the land. Like the great majority of the others, Newton seemed destined to eccentric provincial mediocrity. In the same year as he bought his tu'penny ha'penny notebook, his mother called him home to run the farm. He was just seventeen.

But this time all was not sweetness and light at home. Newton's mind was now afire with something more absorbing than fantasies of pyromania (though it may well have appeared equally disturbed). Psychological explanations of Newton's sudden all-consuming obsession with science abound—from discovering Father's clues, to a demented need for escape into an ordered world free from psychic anxiety. (This complex and often contradictory multiplicity of explanations is useful, if only as a reminder of the complex and often contradictory nature of the unique entity it attempts to describe: Newton's mind.) But one thing is certain, this overwhelming interest in science gripped Newton's adolescent mind with the force of an addiction. And was to retain this force, virtually without ceasing, for *thirty-seven years*.

As a farmer, the seventeen-year-old Newton was worse than useless. Set to watch over the sheep, he would settle in the shade of a tree with a book. When he went to market in Grantham, he left the farmhand to sell the produce and livestock while he nipped around to the house of his former landlord Mr. Clark to pick up some more books (one of Mr. Clark's relatives had left his collection in the attic). The sheep broke loose over the hills, the pigs overran the neighbor's cornfields, and the boundary fences fell into an illegal state of disrepair. As a result, Newton was hauled before the courts and fined four shillings and four pennies. (The cost of a good pair of shoes.) Newton's first recognized qualification was a criminal record.

Mother had no idea what to do, and life at home was fraught. In a "list of sins" which Newton drew up some years later, this period includes such items as: "peevishness with my mother," "falling out with the servants," "refusing to go to the close at my mother's command," and "punching my sister." Like most teenagers, he knew what he didn't want to do. Unlike most teenagers, he knew precisely what he did want to do. Newton continued reading avidly, making models, conducting scientific experiments, calculating and sketching diagrams in his notebook.

Fortunately two people had recognized Newton's exceptional talents. One was John Stokes, his schoolmaster at Grantham; the other was his maternal uncle William Ayscough, rector of the nearby village of Burton Coggles, who happened to be a graduate of Trinity College, Cambridge. Between them, they managed to persuade Newton's mother to send him back to school in Grantham, where Stokes could prepare him for entrance to Trinity College, Cambridge.

Newton returned to live with Mr. Clark the apothecary, where he continued to devour the collection of books, and now began decorating his room with all kinds of drawings. According to Mr. Clark's stepdaughter, he also formed a romantic attachment with her during this period. She was several years younger than he, and the romance appears to have been largely of her own imagining. This is the only occasion in Newton's life when his name was to be romantically linked with a woman.

Newton went up to Cambridge in June 1661, where he was admitted to Trinity College. According to a contemporary historian, Trinity College was at the time "the stateliest and most uniform Colledge in Christendom." Its academic prowess was entering a similar class to its appearance—though Cambridge was still generally seen as lagging behind the great universities of Europe, such as the Sorbonne and Milan. England had not only undergone a political revolution, it was in the process of an intellectual one that was unsurpassed (and as yet largely unrecognized) in Europe. This was to culminate in the works of Newton—but its lesser lights included men of such stature as Harvey (whose discovery of the circulation of the blood ushered in modern medicine), Halley (the great astronomer, after whom the comet is named), Hobbes (the most perceptive political theorist of his era), Locke (whose empiricism changed the course of philosophy, and whose ideas were to shape the

U.S. Constitution), and Boyle (the pioneer chemist).

Newton was eighteen when he arrived at Cambridge, two years older than the average student. He was also much poorer than the average student, and was only admitted as a scholarship student. This required him to act as a form of valet to his tutor. Fortunately, his tutor only deigned to take up residence for five weeks of the year, so Newton had most of his time to himself.

There were few distractions at Cambridge in those days. According to a visiting German traveler, outside the university itself Cambridge was "no better than a village . . . one of the sorriest places in the world." The village taverns were filled with jolly trollops and roistering young gentlefolk. (These were Newton's fellow undergraduates, less than a third of whom bothered to take a degree.) The year prior to Newton's arrival, Charles II had ascended to the throne. After the Puritan excesses of the Commonwealth, the more conspicuous excesses of the Restoration era were now in full swing. But Newton's Puritanism did not depend upon the political climate. Between marathon bouts of studying, which frequently lasted until dawn, Newton might sit in his rooms making lists of his sins (none of which ever rose to roistering or trollops).

Despite the growing intellectual revolution in England, education in the universities remained for the most part heavily rooted in the Aristotelianism of the medieval era. The earth still stood at the center of the universe, which consisted of earth, air, fire, and water. These "elements" were reflected in our four "humours": blood, phlegm, choler, and melancholy, whose "balance" governed our health. And so forth . . . An invincibly coherent world according to its own assumptions, whose inadequacy was only gradually exposed.

The first serious cracks in the edifice of Aristotelianism had begun to appear in Europe during the early seventeenth century. The Polish priest Copernicus had suggested a heliocentric solar system. This had led the German astronomer Kepler, working in

Prague, to propose laws of planetary motion. The Italian physicist Galileo had then put forward a new mechanics based on this (before being forced to recant his views by the Catholic Church). Meanwhile Descartes' philosophy of doubt had shown Aristotelianism, the basis of the Church's scientific teaching, to be devoid of analytic or perceptual justification. Such were the pioneers who stimulated the English intellectual revolution, and the undergraduate Newton was soon becoming heavily influenced by their discoveries.

Of equal importance, Newton also began learning the new mathematics which supported these discoveries, and upon which any future discoveries would need to be based. During the previous century the advances in astronomy and navigation had required new and more refined methods of calculation and exactitude. As a result, mathematics had undergone a revolution that paralleled the recent scientific discoveries. Here too the lineaments of an increasingly precise structure were beginning to emerge from the medieval mists. In 1585 the Flemish civil servant Stevin proposed the decimal system for measurement of amounts less than one; and in the early years of the seventeenth century the Scottish baron Napier invented logarithms. This mathematical revolution had come to full flower in France. Here three of the greatest mathematicians of all time—Descartes, Fermat, and Pascal—had all reached the height of their powers by the mid-seventeenth century.

During Newton's undergraduate period he studied and absorbed the lessons of Descartes (though how much he knew of Pascal and Fermat remains an open question). Descartes had invented Cartesian coordinates (named after him): the three axes that enabled every geometric point (or straight line, or curve, or shape) in space to be precisely mapped. Algebra was also introduced into geometry, liberating it from the particularity of arithmetic, and analytic geometry was born. A curve could now be represented by an equation, as in the two-axis figure below.

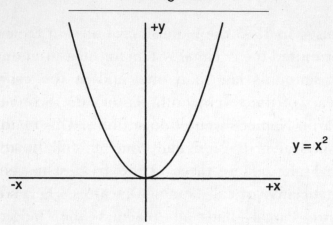

$$y = x^2$$

More significantly, Descartes' mathematics (and his philosophy) viewed the world as a vast intricate mechanical apparatus. Previously, Aristotelianism had viewed the world in terms of qualities (earth, air, etc); now it was viewed in terms of quantities—which could be measured.

Newton began keeping a notebook, entitled "Certain Philosophical Questions," (*Quaestiones quaedam philosophicae*) with a heading "My best friend is truth." In this we see him absorbing Descartes' proposal that all reality consists of particles of matter in motion, and that all natural phenomena come about through the interaction of these particles. The French scientist-philosopher Gassendi revived the ancient Greek notion that these ultimate particles were discrete indestructible atoms. Newton's notebooks show that he was also aware of parallel developments by Boyle, whose experiments were beginning to suggest the existence of basic chemical elements.

Boyle's discoveries laid the groundwork for Newton's future work in chemistry, but Newton's interest in chemistry was hardly forward looking. In the seventeenth century chemistry was only just beginning to emerge from the mumbo-jumbo of alchemy. In pursuit of his chemical interests, Newton also began reading works on alchemy, magic, and the hermetic tradition. These purported to explain natural phenomena in terms of metaphysical gobbledegook.

It would be nice to think that Newton read all this nonsense by way of light amusement, after the extremely exacting rigors of his mathematical and scientific studies. But this was not the case. Newton took his alchemy seriously. This was on a par with his obsessive study of the Bible. Here too, he might find clues as to Father's identity.

These two ways of looking at the world—the physical and the metaphysical—would seem to us mere mortals to be mutually exclusive. But they were not for Newton. Indeed, the contradiction between these two world views appears to have acted in some ways as a stimulant to his mental processes.

But Newton's discoveries were not all fool's gold. By the time he had been an undergraduate for three years, Newton was already making important mathematical discoveries. He had worked out the binomial theorem for fractions. The binomial theorem involves the formula for the expansion of a binomial expression. That is, one containing the sum of two variables, such as $(x + y)$, raised to a given power n—expressed $(x + y)^n$.

A simple example shows:

$$(x + y)^2 = x^2 + 2xy + y^2$$

But when n (the power) is not a whole number, the expansion becomes an infinite series.

An example with a single variable is as follows:

$$(1 + x)^{1/2} = 1 + \frac{1}{2}x - \frac{1}{8}x^2 + \frac{1}{16}x^3 + \ldots$$

and so on, ad infinitum.

Newton worked out a general rule for such expansions. As we shall see, this work on infinite series was Newton's first step toward one of the greatest mathematical discoveries of all time: calculus.

Newton received his B.A. in June 1665. Professor Barrow, his examiner, formed "an indifferent opinion" of his abilities: Newton didn't even know his basic Euclid. Newton had indeed sorely neglected the syllabus. What Professor Barrow didn't realize was that Newton was already advancing beyond Descartes, who in his turn had already advanced beyond Euclid. Newton was almost entirely self-taught—in the sense that he worked largely alone, from books. All his truly amazing work was confined to his notebooks—which nobody else had seen. Despite the gaps in his knowledge, Newton was allowed to continue studying for an M.A. (Students who actually studied were evidently a rarity in need of preservation.)

Newton seemed to thrive in isolation, and events now conspired to make sure this continued. Late in 1664 two French sailors had been found in the London slums around Drury Lane dying of the bubonic plague. The disease quickly spread throughout the city (it was eventually to cause over eighty thousand deaths in London alone), and then out along the stagecoach routes and cattle drovers' trails into the country at large. All who could began to flee the centers of population, and by August 1665 Cambridge University had effectively closed down. Newton returned to Woolsthorpe, where he remained for around a year.

This was to result in an *annus mirabilis* the

like of which has never been seen in science before or since. (The only near competitor was Einstein's 1905, when he discovered the special theory of relativity, where he proposed that light consisted of quanta and provided the molecular explanation of Brownian movement.)

Newton's first major breakthrough was the development of calculus. The ability to represent an algebraic formula on a graph now meant that certain algebraic problems were susceptible to geometric solutions. For instance, the rate of change of x against y for any given values (i.e., at any given point on a curve) is the tangent to the curve at that point.

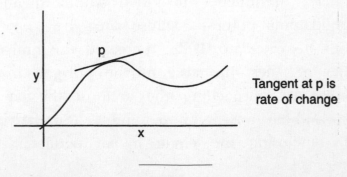

Tangent at p is
rate of change

It's easy enough to find the tangent to a circle, or even to a regular curve, but how do we find it for a variable curve? This is done by calculus. The basic notion of calculus was discovered by the ancient Greeks.

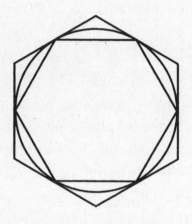

In order to determine the area of a circle, Archimedes inscribed within it an equilateral polygon, whose area he knew how to calculate. If he increased the number of sides, the area of the polygon increased, approaching the area of the circle—which was its upper limit. (Similarly, by enclosing the circle in an equilateral polygon he could discover the lower limit, the answer lying between the two limits.)

However, if the sides of the polygon were increased to infinite, this also would give the area of the circle.

This principle could similarly be applied to the tangent to a curve.

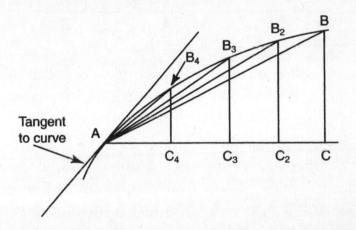

As the distance between A and B becomes infinitely small, tending toward its limit zero, the side AB opposite the right angle ACB approximates closer to the tangent, becoming this at the limit.

The same principle can be applied in calculating the area beneath a curve.

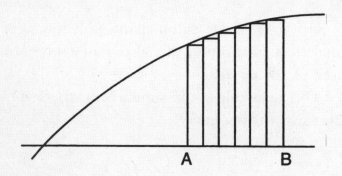

As the number of rectangles between A and B is increased toward infinite, their area approaches the area under the curve as their limit.

Once again, this was a problem with an infinite series arising from functions with two variables—just as Newton had dealt with in binomial theorem. To begin with, the calculations involved in such problems were seen as involving static sums of infinitely small quantities. Newton's great insight was to see this problem instead as one of *mobility*—treating the curve not as a static object, but as the locus of a *moving point*. (Indicatively Newton first called his method "fluxions," implying flow—not calculus, as it was later to be known.) His innovation was thus to introduce the notion of time.

Newton's method for finding the tangent to a point on a curve is now known as differential calculus. This regards the ever-varying quantity of a *moving point* as if it were made up of an infinitely large number of infinitely tiny changes.

For instance, the velocity *(v)* of a body at a particular instant is seen as the infinitesimally small distance it covers *(ds)* in the infinitesimally small, decreasing to zero, amount of time *(dt)*. Therefore

$$v = \frac{ds}{dt}$$

Now as dt $\rightarrow$ 0 (i.e., reaches its limit at zero) so *v* reaches the limit which is its exact speed at the given moment.

Fortunately Newton's long and mind-bogglingly complex calculations eventually yielded an easily applicable rule of thumb—as used by all budding mathematicians who know what to do, but don't really know what they're doing. ("Just follow the rule, young man.") Put in the very simplest terms, for the formula

$$y = x^n$$

the derivative $\dfrac{dy}{dx} = n\ x^{n-1}$

For example: make $n = 2$, thus $y = x^2$

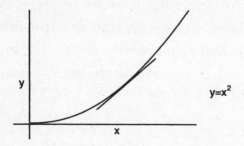

The rate of change at *any* point, $\dfrac{dx}{dy}$, equals $2x^{2-1} = 2x$

In other words, the gradient of the tangent at any point on the curve will always be 2x.

This process of differential calculus provided the new mathematics with one of its most powerful tools—allowing the calculation of all kinds of rates of change. This included for instance the maximum and minimum points in any curve—which occur when the gradient, or the rate of change, $\frac{dx}{dy}$, is equal to zero:

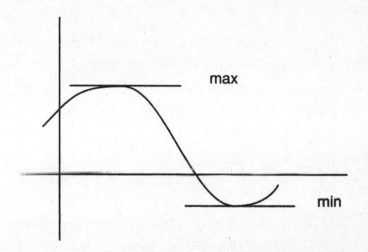

Newton then went on to extend his method of fluxions to include what is now known as integral calculus. This is essentially the reverse technique to differential calculus, and is used for calculating the area beneath a curve.

For example, the velocity of a point *(v)* can be expressed in terms of the infinitely small distance *ds* traveled in the brief moment *dt*. Thus:

$$ds = v\ dt$$

The measurable distance *s* which the point travels between time t_1 and t_2 is found by continuously summing the changes in this interval, which is known as integration.

This is expressed:

$$s = \int_{t_2}^{t_1} vdt \quad \left(\int \text{ is the sign for integration} \right)$$

Greatly simplified, this involves applying the opposite technique to integration. So, for the same formula

$$y = x^n$$

instead of $\dfrac{dy}{dx} = n\ x^{n-1}$

for integration we get $\int x^n \, dx = \dfrac{x^{n+1}}{n+1}$

This immensely useful technique could be used for such problems as finding the area of any kind of shape described by a formulaic curve (and rotating about the axis produced a volume). It could also be used for any problem where continuous summing of infinitesimal changes was required.

At this stage, Newton's calculus still remained in embryo form. But even so, he now had the technique that enabled him to undertake his major work. Newton's transcendent achievement during the course of 1665–66 was of course concerning gravity.

Newton was later asked how he had achieved this and his other epoch-making discoveries. "By always thinking unto them," he replied. "I keep the subject constantly before me and wait until the first dawnings open little by little into the full light." According to the famous story, the "first dawnings" of his theory of gravity came to Newton when he saw an apple drop from a tree. This is often dismissed as sheer legend. But according to Newton's early biographer, Stukely: "he told me . . . the notion of gravitation came into his mind . . . occasion'd by the fall of an apple, as he sat in contemplative mood."

It is important to understand the full significance of what Newton understood at that moment. What did he know already, and what did his eventual theory of gravity explain?

The key to it all was Kepler—who had taken over twenty years of painstaking observation and endless calculation before arriving at his three laws of planetary motion. These were published in 1609 and stated: one, the planets travel in ellipses around the sun, and the sun is at one focus of these eliptical orbits; two, a straight line joining the sun and a planet sweeps out equal areas in equal times; (in diagram below: time taken for planet p to travel from a to b, is the same as from c to d, and the area x is equal to area y); three, the square of the time taken for one complete orbit by a planet is proportional to the cube of its average distance from the sun; (in the diagram if planet p takes time T to complete an orbit, and r is the average radius of this orbit, then: $T^2 = r^3$).

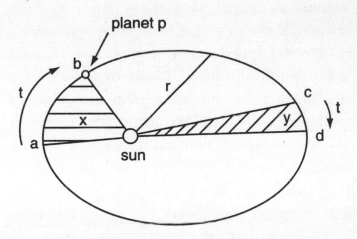

T = time taken to complete one orbit

Meanwhile back on earth, Galileo had confirmed by experiments, said to have been conducted from the Leaning Tower of Pisa, that a falling body accelerates at a uniform rate. He also derived a formula for the parabolic path of a projectile.

ton's genius was to put Kepler's laws and Galileo's findings together. The notion of gravity that came to him when the apple fell from the tree would eventually be seen as the same power that held the moon in orbit around the earth, and the planets in orbit around the sun. The laws that applied on earth also applied to heavenly bodies. This was a stupendous intuition. At one step our understanding was no longer earthbound, but extended throughout the universe. (Kepler's laws merely described what happened; Newton explained *why*.)

Newton was not to publish his ideas for over twenty years. There are several reasons for this. At first he only regarded gravity as applying on earth. Later, when he extended this to extraterrestrial bodies, he couldn't quite work out the mathematics of it. How did the earth's gravitational force actually work? Did it attract the moon from its center or from its surface, or from somewhere in between? Not until he had refined the techniques of his newly discovered calculus was he able to overcome such problems. Yet these weren't the only reasons for his silence.

Some have called Newton a secretive character. But this is not strictly true. The fact is, Newton couldn't abide being contradicted, even in the most trivial matter. It was liable to make him burst into one of his uncontrollable rages. So rather than face the questioning of his fellow scientists, he preferred to keep his discoveries to himself.

Needless to say, such psychology only accounts for Newton's personality. It may be likened to a map of the world. This outlines the contours and shapes, but in no way explains the magnificence and profusion of the reality. The sheer quality of Newton's mind remains utterly inexplicable.

During the twenty years before Newton published his findings on gravity, his initial insight became refined into a comprehensive system. It was this that finally appeared in his masterpiece, the *Principia*. Here Newton went one step further than Kepler and Galileo, putting forward three laws of his own which superseded their findings.

Newton's first law of motion posits a theory of inertia, stating that a body remains at rest or in uniform motion along a straight line unless it is acted upon by an outside force. Things moved through space because there was nothing to stop them after they had initially been set in motion. For the first time, the movement of bodies through the heavens was *explained*—without recourse to divine juggling or locomotion by angels. (Though not until three centuries later did the big bang theory explain how this initial motion came into being.)

Newton's second law of motion states that the rate of change of momentum (mass × velocity) of a moving body is proportional to the force impressed upon it. In other words, the affect of a continuous force upon a stationary body or one in uniform motion is to make it *accelerate*. Galileo had discovered this when he dropped objects from the Leaning Tower of Pisa. The pull of gravity makes a body accelerate. The same happens when the moon orbits the earth.

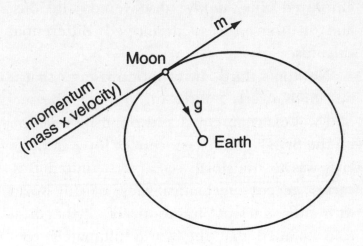

The continuously acting force of gravity (g) impels the moon to accelerate toward the earth, but the moon's momentum (mass × velocity) impels it along the line of force m. The resulting continuous balance of forces keeps it in orbit. To discover the force of gravity operating here, Newton had to calculate the moon's rate of change of momentum. As the moon's orbit is an irregular ellipse, this involved calculating the velocity of an object moving in a curve. It was in his earliest attempts to solve this problem that Newton employed his newly discovered fluxions, and in the process developed differential calculus.

Newton's third law of motion states that if one body exerts a force on another, the second will exert an equal and opposite reaction on the first. Newton's concept of force in these laws was to transform science. It united Descartes' recent mechanical view of the world with the ancient tradition of Pythagoras, who claimed that the world ultimately consisted of numbers. This combining of mechanics and mathematics not only explained how the world worked, but meant we could also calculate precisely what was happening in it.

By using these three fundamental laws, Newton was finally able to conclude how gravitational force acted between two bodies. He showed that this is directly proportional to the product of their two masses and inversely proportional to the square of the distance between their two centers. This was expressed in his celebrated formula (the $e = mc^2$ of its day)

$$F = \frac{m_1 m_2 G}{d^2}$$

where F is the force of gravitational attraction, m_1 and m_2 are the masses of the earth and the moon, d is the distance between their centers, and G is the gravitational constant. What had set him on the path to this formula was the possibility of the inverse square relation—and this may well have been the original realization provoked by the falling apple. Newton didn't understand the entire notion of gravity in a flash; but this was what set him on the long and complex mathematical journey that ended in his law of gravitation. Even so, it was to be a century before the eccentric English physicist Cavendish managed to determine the value of G, the gravitational constant. However, this incompleteness didn't stop Newton from making sweeping claims for his new law. He asserted that the law of gravitation applied throughout the universe. This was, of course, a hypothesis: Newton's calculations were based entirely on observations of the moon and the discovered planets. But Newton would brook no objection, famously claiming: *"Hypotheses non fingo."* (I do not make up hypotheses.)

It is difficult for us to understand the sheer flimsiness of Newton's claim with regard to his gravity law—which he insisted upon calling the *universal* law of gravity. One of the greatest human insights of all time was in fact little more than a hunch—a guess of transcendent genius. The twentieth-century mathematician and philosopher Whitehead provided a useful corrective, both to Newton and us all: ''The pathetic desire of mankind to find themselves starting from an intellectual basis which is clear, distinct and certain, is illustrated by Newton's boast *hypotheses non fingo,* at the same time when he enunciated his law of universal gravitation. This law states that every particle of matter attracts every other particle of matter, though at the moment of enunciation only planets and heavenly bodies had been observed to attract 'particles of matter.' ''

There may have been scanty *scientific* evidence for Newton's claim of universality, but it certainly accounted for many observed facts and eccentricities of planetary movement. Most interesting, it *explained* Kepler's laws and also accounted for irregularities in the orbits of the moon and planets (these occurred when they were affected by the gravitational pull of other passing planets as well as that of the sun).

Newton's bold guess changed everything. From then on scientists believed that anything that happened in the universe could be explained in terms of mathematics. This has remained one of the central beliefs of modern science. Indeed, it is the cornerstone of the continuing belief in an ultimate theory that will explain the fundamental workings of the universe and everything in it.

The third momentous discovery Newton made during his *annus mirabilis* at Woolsthorpe was concerning light. Previously it had been thought that color was created by a mixture of light and darkness. Newton realized that this was not supported by experimental evidence. The printed page of a book—which contained both white and black—did not appear colored when viewed from a distance so that the two blended. It appeared gray.

Newton conducted a number of experiments at home in his darkened room with a glass prism. When he let in a chink of daylight between the curtains so that a ray of white light passed through a prism, it was refracted (bent by the glass). But different parts of the beam were refracted by different amounts, and the beam emerged split into colors. These colors were the same, and in the same order, as they appeared in the rainbow: red, orange, yellow, green, blue, indigo, violet.

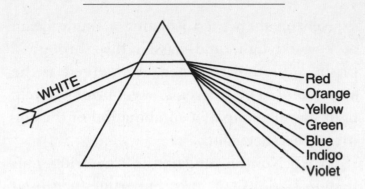

Were these colors somehow created by the clear glass of the prism? Newton now passed the rainbow beam of light through a farther prism, this time upside down. The beams of colored light then reconverged and emerged as a single beam of white light.

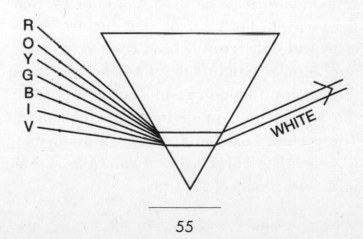

Newton then tried isolating a single beam of colored light and passed this through a prism. Although refracted, it emerged as the same color. The inference was obvious. White light was made up of a combination of the colors of the spectrum.

When Newton returned to Cambridge, his demonstrations of this experiment caused quite a stir, and he was elected a fellow of Trinity College. But he remained reticent about his other discoveries. Calculus and his ideas on gravity were still in the early stages, and he had no wish to enter into discussions on these matters. (Someone might have had the temerity to contradict him.) However, he did make an exception in the case of his former tutor Barrow, the Lucasian Professor of Mathematics. (This chair had only recently been created: an early indication of Cambridge's diversification from the classical tradition and its emergence from the stranglehold of defunct Aristotelianism. The present Lucasian Professor of Mathematics is Stephen Hawking, best known as author of *A Brief History of Time*.)

Barrow was an exceptional man, in many ways the very opposite of Newton. No stereotype mathematician—he was an affable fellow of fine physique who enjoyed boxing and had traveled as far as Constantinople (where he had won a wrestling contest). But Barrow had two points that appealed to Newton: He was a deeply religious man and a fine mathematician. Indeed, Barrow's understanding of the latest mathematical advances undoubtedly aided Newton in his development of calculus.

Unusual for a young man in his situation, Newton was not drawn toward father figures. (Father was in heaven.) A brief exception appears to have been Isaac Barrow (whose first name even echoed that of Newton's father). For once, Newton had found someone he could admire. Like Newton, Barrow worked long hours and slept little. Newton must also have recognized a similar self-absorption in Barrow's absentmindedness, for according to one of Barrow's contemporaries, he was "most negligent in his dresse . . . like the veriest scholar that ever I mett with." Newton presented a similar picture: "he would go very carelessly, wth Shooes down at Heels, Stockins unty'd, surplice on, & his Head scarcely comb'd." The professor and his twenty-four-year-old colleague must have made a fine pair, amid the bewigged Restoration fops.

Barrow appears to have been the only man in Cambridge who realized the truly exceptional extent of Newton's abilities. Outside the university, of course, no one had even heard of him. Yet by now Newton had made discoveries that placed him far in advance of any scientist or mathematician alive.

Barrow is said to have treated Newton as a son, giving him presents on his birthday. Though drawn to Barrow, Newton's attitude was more wary. His mother's early desertion had left him profoundly ambivalent toward the few who would manage to penetrate his carapace of otherworldly indifference.

In 1669 Barrow resigned as Lucasian Professor of Mathematics in order to pursue theological studies. He made sure that Newton was appointed in his place. The Lucasian professor was expected to become a member of the clergy, but Barrow interceded on Newton's behalf and Newton was not required to take up holy orders. This indicates that Barrow was at least partly aware of Newton's less orthodox researches.

Alongside his scientific advances, Newton had also made astonishing advances in his biblical studies. While reading the earliest versions of the New Testament in their original languages, he had become convinced that these texts had been corrupted by later translators and commentators for their own purposes. The idea of the Trinity (Father, Son, and Holy Ghost) was a complete hoax, a fraudulent conception foisted on Christianity by scheming deviants. Christ had not been divine, and we should pray directly to God the Father.

Such beliefs had been declared heretical by the Council of Nicea in A.D. 325. Likewise, the authorities of Trinity would not have welcomed the news that their college was named after a theological hoax—but this did not deter Newton. However, in keeping with his usual practice, he confined his findings to his notebooks.

In this case there *was* a secretive element to Newton's behavior. Religion was taken very seriously indeed in seventeenth-century England—the civil war, the persecution of heretics, fear of Catholicism, distaste for Puritanism, and more, had produced a dangerous cocktail of prejudices. Not surprisingly, Newton developed a paranoid fear of being exposed as a heretic, and this was to last until the end of his days. But his profound belief in God, coupled with his unconscious need to communicate with Father directly, meant that he couldn't stop himself. Newton's impulse toward truth was as strong in religion as it was in science: Here too he was searching for God's clues.

As if all this wasn't enough, Newton also continued with his alchemical researches in the hermetic tradition. The Lucasian Professor of

Mathematics even went so far as to have a furnace constructed in the garden outside his college rooms, so that he could carry out alchemical experiments. His activities here were presumably passed off to the college authorities as chemistry, but one glance at Newton's notebooks makes it clear that his interest was in transmuting base metals into gold.

Incredibly, the first scientist of his age was also one of its leading sorcerers. As the twentieth-century economist Keynes pointed out: Far from being a modern man of the new scientific era, Newton was in fact the last of the great Renaissance magicians, "the last wonderchild to whom the Magi could do sincere and appropriate homage."

But was Newton really just wasting his time (and his prodigious talents) on balderdash? Concurring with sane opinion, Keynes found himself forced to dismiss Newton's alchemy as "wholly devoid of scientific value."

Alas, the truth is otherwise. Though it is unlikely that the Magic Circle will produce the next Einstein, there is no denying that alchemy played a central role in shaping Newton's *scientific* ideas. The evidence is unfortunately compelling. As we have seen, earlier in the century Descartes had proposed a purely mechanical explanation of the world. But as the scientific revolution progressed, a number of the new British scientists had begun to suspect that the world worked in a more complex fashion than the interior of a watch. In the view of the chemist Boyle (who was also an avid alchemist on the quiet), mechanics was not adequate to explain several natural phenomena occurring in chemistry and biology.

Newton conceived of the idea that such events were the result of an active principle, which complemented Descartes' mechanical inertia principle. This active principle was due to "occult qualities . . . incapable of being discovered and made manifest." From here it was but a short step to the notion of "force," Newton's idea that was to transform the whole of science. It may be difficult for us to swallow, but the central revolutionary concept of Newton's laws of motion originated in magic. As

Leonardo, that other great Renaissance magi, observed: "There is more in the world than ever man will understand."

Newton's alchemical activities also assisted him in other ways. Alchemy may have lacked demonstrable results (i.e., no gold), but its methods involved considerable ingenuity—not least in the assembly of apparatus. (Much of this experimental expertise was adopted piece-meal by the embryonic science of chemistry, which as a result got off to a flying start.) Newton had displayed exceptional practical exper-tise during his youth in Grantham, yet he had found less occasion for using this talent during his scientific and mathematical re-searches at Cambridge and Woolsthorpe. But for alchemy, he might even have avoided com-plex experimental endeavors. As it was, al-chemy refined his practical abilities, which gave him confidence to seek practical solu-tions. The best-known example of this is his telescope.

As we have seen, when light passes through a prism, it produces a spectrum. A similar side effect takes place with lenses, which are liable to produce images with colored fringes. This was beginning to hamper the effectiveness of the increasingly large telescopes being used in astronomy (the nuclear physics of the sixteenth and seventeenth centuries, which opened up the new scientific era). By the turn of the seventeenth century, telescopes over two hundred feet high were being built, but their magnified images were more and more subject to color interference, known as "chromatic aberration." It looked as if telescopes had reached their limit, bringing an end to further astronomical discovery.

Newton first attempted to solve this problem by grinding different-shaped lenses, but this proved unsuccessful. Once again he concentrated his mind on the problems involved: "always thinking unto them," both night and day, until eventually he saw the answer. Newton's solution to the telescope problem was a classic "stroke of genius." It was so simple and effective that it transformed telescopes forever. Instead of concentrating the final image by refraction through a lens, he did this with a parabolic mirror.

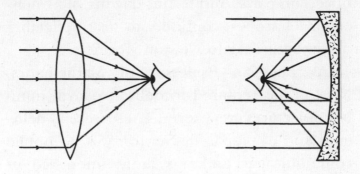

Reflection eliminated "chromatic aberration" as the light did not pass through the glass, it merely bounced off it. And there was a further advantage to this method: As the light didn't pass through glass, none of it was absorbed. (This was vital in the observation of smaller distant bodies, such as the moons of Jupiter, which only reflected small amounts of light.) Newton's method also made the telescope much smaller. The first telescope he produced was just six inches long and one inch in diameter—yet it had a magnification of over thirty times. Newton built this telescope entirely by himself, even going so far as to construct his own tools for the manufacture of certain parts. And to this day the most powerful telescopes continue to use reflecting dishes according to Newton's principle.

As Lucasian Professor of Mathematics, Newton was required to deliver a small number of lectures every term. These were ill delivered and ill prepared. Newton was no public communicator, and was more interested in talking about his continuing researches rather than final results. After the initial clarity and excitement of the prism demonstration showing how white light is composed of color, a fog of abstruse speculation and theory descended. In the end, Newton was often left muttering to himself in an empty lecture hall.

But this all changed when he produced his new telescope in 1668. Newton was so proud of his handiwork that for once he couldn't resist showing it off. Word began to spread through Cambridge, and eventually leaked out farther afield. The Royal Society in London heard of this "wondrous Instrument" and asked to see it. By now Newton was building a second larger version, which was nine inches long by two inches in diameter. In 1671 Barrow took this to London for him, where it caused such a sensation that it was demonstrated to Charles II. As a result, Newton was elected a member of the Royal Society (which had been founded in 1660, and was now the leading scientific society in Europe). Newton at last had contact with the finest minds in the British scientific revolution.

Encouraged by this honor, Newton was persuaded to divulge some of his secrets. In 1672 he sent the Royal Society a paper on optics, describing his theory of light and color. In the words of the secretary of the Royal Society, this paper "mett both with a singular attention and an uncommon applause," though a few members demurred. Among these was the cantankerous physicist Hooke, one of the few scientists of sufficient caliber to contradict Newton. Hooke was an experimenter of genius, but seldom carried his insights through to fruition. He proposed an early but ill-developed wave theory of light; he anticipated the steam engine, but in an impractical form; and his pioneering microscopic observations led him to invent the term "cell" (though he applied it wrongly). Hooke had also undertaken experiments on light, using prisms, and had come up with a characteristically cockeyed theory of his own. Hooke was a leading power at the Royal Society, and he considered optics to be his territory. He wrote a patronizing critique of Newton's paper pooh-poohing his conclusions. Letters were exchanged, and when Newton published a second paper elaborating his findings Hooke accused him of plagiarism.

Newton was psychologically incapable of accepting criticism at the best of times. As a result of Hooke's accusation he was unable to contain himself, and his rage knew no bounds. Hooke became his sworn enemy—a role the irascible Hooke was only too pleased to assume. But this was no passing fury on Newton's behalf. He was so upset by this incident that his work suffered for over two years. He acrimoniously resigned from the Royal Society (an unheard of act, which fortunately was not accepted), and he swore that he would publish no further scientific work.

But still the controversy dragged on. Newton became involved in correspondence with some English Jesuits in Liège, who sought clarification of his original prism experiment. Their experiments had not produced the same effect which Newton claimed. This correspondence was to drag on for some years, with the Jesuits eventually denying the veracity of Newton's results. Already considerably agitated by this correspondence, Newton now mistook stupidity for conspiracy. His paroxysm became so great that he swore to abandon science: "I will resolutely bid adew to it eternally"—and suffered a complete nervous breakdown.

While he recovered, Newton was as good as his word. He abandoned his scientific researches and buried himself in biblical and hermetic studies. The entire notion of the Trinity was rigged, supported only by fourth-century documents forged by St. Athanasius. Using the star regulus and copper it was possible magically to produce the hermaphrodite known as the net, which consisted of the male seed of Mars and the female principle of Venus. And so forth. (Newton's library consisted of over one hundred forty books on alchemy alone; and according to a biographer, at his death his papers contained "half a million worthless words on chemistry.")

In 1679 Newton's mother died. He had been deserted by her for the last time: He was alone. It was Freud who first noted that the greatest intellectual insights often come to their discoverers after they have suffered a profound loss. And this was to be no exception. As a result of his occult notions concerning attraction and repulsion, Newton had conceived of the idea of forces. But so far he had applied this notion only to earthbound phenomena. He now received a letter from Hooke, who wanted to patch up their differences. Hooke informed Newton about his analysis of planetary motion, including his idea of a central attraction that kept the planets in elliptical orbits. Hooke had guessed that this probably worked according to an inverse square law—unaware that Newton had worked out the mathematics of an inverse square law a decade or so previously.

Newton refused to be drawn into a lengthy correspondence with Hooke, but later admitted that Hooke's letter prompted him to apply his own inverse square law—whose mathematics was firmly based on Kepler's third law—to the elliptical orbits of the planets. It was this that in turn prompted Newton to one of his greatest insights. His notion of force—so far seen only in terms of terrestrial phenomena— also applied to orbital mechanics.

Newton was now on the brink of the concept of universal gravitation. But it was to be five years before the penny dropped. And once again this was prompted by his old bête noire, Hooke.

In 1684 the odious Hooke bragged to the

astronomer Halley that there was no longer any problem over planetary motion. He himself had worked out an inverse square law that governed the motion of heavenly bodies. Halley remained unconvinced by Hooke's explanation, which had little mathematical backing.

Halley decided to consult Newton, who informed him that he had produced a theory of orbital dynamics some years earlier, and had the calculations to back it up. Halley managed to persuade Newton to send him a paper outlining his findings. Newton duly began a paper entitled "On the Motion of Heavenly Bodies in Orbit" *(De motu corporum in gyrum)*, which Halley received seven months later.

Halley was so impressed by this paper that he traveled once more to Cambridge, where he discovered that Newton had a vast store of unpublished papers. Halley's father had recently been murdered, leaving him a fortune. Halley suggested that he would be willing to finance the publication of Newton's papers.

Meanwhile the writing of *De motu* had caused Newton to think beyond his ideas of planetary motion, and he had finally conceived of the idea of universal gravitation. In the flush of inspiration he agreed to Halley's scheme, and settled down to the necessary calculations.

For two and a half years Newton worked in isolation, producing the work that was to be his masterpiece, "The Mathematical Principles of Natural Philosophy" *(Philosophiae naturalis principia mathematica)*. Nowadays this book, generally acknowledged as the finest work of science ever produced, is usually referred to simply as the *Principia*.

In keeping with the medieval custom that still prevailed, the *Principia* was written in Latin (which continued to serve as the international language). Its full title stems from the fact that in those days science was still considered a branch of philosophy, and was referred to as natural philosophy—though the implications of Newton's revolutionary work were to extend into philosophy itself. From now on, no philosopher could ignore this new cosmology, based upon experiment. It was no longer possible to explain the world by simply thinking about it and producing abstract principles. Concrete experience had to be taken into account. Heavily influenced by Newton's *scientific* discoveries, the philosopher Locke was to produce empiricism, which states that our knowledge derives fundamentally from experience, thus laying the foundations of modern philosophy. Newton's *Principia* was to transform the entire way we thought about the world.

This work was also seen as one of the seminal works of the Age of Reason. An atmosphere of intellectual optimism began to prevail—science had shown that the world was constructed according to basic principles that could be elaborated according to reason. All could be known: Science held the key to life, the universe, and everything.

Yet paradoxically this most modern of books was not only written in Latin, but set out in the style of the ancient Greeks. Its three laws of motion and law of universal gravitation may have been the cornerstones of modern science, but these were set down and proved by geometrical reasoning, just as Euclid had done two thousand years previously. Newton was aware that he was writing a classic, and he wanted it to be in the classic style. It would have been much easier (and made much more sense) for him to have used his new discovery, calculus. But this new method, which was capable of transforming mathematics, he preferred to keep a secret. (There is only a passing reference to it in the *Principia;* but as we shall see, this was to establish a vital precedent.)

The manuscript of the *Principia* was first sent to the Royal Society, whose secretary was now Hooke. As soon as Hooke read Newton's manuscript, he accused Newton of plagiarism. He had written to Newton six years previously divulging his inverse square law. Newton had based his work on stolen property.

The effect of this accusation was predictable. Newton was unable to contain his rage. He had discovered the inverse square law, and worked out its mathematics, ten years before Hooke's letter. But the trouble was, Newton had kept his discovery to himself. Halley appealed to Newton's better nature. Hooke was ill and aging, his antisocial behavior having reduced him to penury. All Hooke really required was some kind of acknowledgment—it would cost Newton nothing to make such a gesture in his *Principia*.

But Halley had underestimated Newton. He had no better nature—and his anger knew no bounds. Instead of inserting an acknowledgment into his work, Newton now vindictively searched through it, eradicating all reference he could find to Hooke—though in the heat of the moment he did manage to miss a few. (Newton's fury was to be no passing matter. As long as Hooke remained secretary of the Royal Society, Newton refused to accept any post in it, refused to let it publish his works, and kept all his manuscripts to himself. Hooke was to linger on for seventeen years, and this state of affairs was resolved only by his death in 1703.)

When Newton's *Principia* was finally published in 1687, it caused a sensation. Newton became internationally famous. Though his concept of "force" was not generally accepted on the Continent, the leading scientists of the day soon recognized him as a worthy successor to Galileo and Descartes.

Meanwhile, James II had introduced a campaign to transform Cambridge into a stronghold of Catholic learning. The academic staff resisted, and the author of the *Principia* became their unlikely champion. Newton, the covert heretic, at last had a legitimate target on which to vent his anxieties. His determined resistance to the king appeared brave and even reckless, though it was driven by forces none suspected. Indeed, but for the flight of James II and the installation of a Protestant monarchy in the form of William and Mary, Newton would have found himself in serious danger.

In recognition of his stand, Newton was appointed a Member of Parliament for the university (which returned three unelected candidates during this period). Being an MP involved periods of residence in London. Here Newton found himself regarded with some awe, and was even invited to dine with the king. He was acknowledged as "the finest of all thinkers" by Locke, whose acquaintance he made—along with such worthies as Wren (who was at the time completing St. Paul's Cathedral), Pepys (the diarist and naval administrator, who had incongruously become president of the Royal Society), and Charles Montague (the ambitious politician who later became Lord Halifax). He also attracted a wide following among the younger generation of scientists, using his growing influence to have several appointed to the few paid university posts open to "natural philosophers."

Among these acolytes was one Fatio dc Duillier, a young Swiss mathematician who had met the German philosopher-mathematician Leibnitz and the Dutch physicist Huygens, inventor of the first genuinely accurate chronometer. Newton took an immediate liking to Fatio, and had soon formed a close emotional attachment to him. Generous references to Fatio even began appearing in Newton's scientific papers, acknowledging snippets of information that Fatio had passed on to him (a rare honor indeed). Newton took lodgings close to Fatio while in London, and Fatio even suggested that he should abandon living in Cambridge altogether and take up a post in London. According to Richard S. Westfall, Newton's great modern biographer, his relationship with Fatio "was the most profound experience of his adult life." When separated, they exchanged increasingly intense letters.

Falling in love (even if Newton remained unaware that this was what had happened to him) gave the forty-eight-year-old scientist renewed energies. In addition to his genuine scientific work in optics, he flung himself into his alchemical pursuits with renewed enthusiasm. According to his assistant, he continued "about six weeks in his Elaboratory, the Fire scarcely going out either Night or day, he sitting up one Night, as I did another, till he had finished his Chymical Experiments." At the same time, his increasing self-confidence emboldened him to write a paper explaining his religious "findings." He even showed this to

Locke, who agreed to have it published anonymously in Holland. But at the last moment Newton got cold feet: Disapproving the Trinity might do irreparable harm to his college (not least, his continuing presence in it).

According to a legendary story, Newton almost caused rather more serious damage to his college. After working through the night, he set off one morning for church, absent-mindedly leaving a candle still burning on his desk. During his absence, this was knocked over by his dog, Diamond. In the ensuing conflagration, years of Newton's priceless unpublished work went up in flames. Upon returning from church, Newton is said to have exclaimed: "Oh, Diamond! Diamond! Thou little knowest the mischief done."

The pressure on Newton's already overtaxed mind became increasingly harmful. Toward the end of 1692 Newton appears to have undergone a crisis of faith in alchemy, which affected him deeply. At the same time another crisis was unfolding. Fatio had been seriously ill. Then suddenly he announced that his mother had died and he would have to return to Switzerland. Newton was distraught, dispatching anguished letters to Fatio, begging him to move to Cambridge with him. Fatio prevaricated, deeply drawn to Newton. The exchange of letters reached fever pitch. And then suddenly stopped. We can only guess why.

Around this time, a fellow don noted that Newton suffered from "a distemper that much seized his head, and left him awake for about five nights altogether." The next four months are shrouded in silence. This is broken by a letter to Pepys, in which Newton informed him: "I am extremely troubled at the embroilment I am in . . . nor have my former consistency of mind." Three days later Locke received a scrawled ink-blotched letter from Newton, written at the Bull Tavern in Shoreditch, east London. In this he begs Locke's forgiveness for "being of the opinion that you endeavoured to embroil me with woemen [and saying] 'twere better if you were dead."

Newton had suffered another mental collapse, from which it took him nearly two years to recover. (Fatio seemingly suffered from an even worse breakdown, disappeared from the mathematical scene altogether, and was next heard of living with an extremist religious sect of French exiles.)

Newton was never again to undertake major scientific work—though he did produce summaries of previously unpublished work, which contributed considerably to his reputation. When Newton had recovered from his illness, his friends encouraged him to seek some prestigious post in London. (He obdurately refused the presidency of the Royal Society while Hooke was still hanging on as secretary.) Newton approached his political pal Montague, and was appointed Warden of the Mint, at the vast salary of two thousand pounds (at the time a skilled laborer was lucky to earn twenty pounds a year).

The appointment of Newton to the Mint was intended as a well-earned sinecure: a reward for England's noblest intellectual ornament. Or so the official story goes. But according to his French admirer Voltaire: "I supposed that the Court and the city of London named him Master of the Mint by acclamation. No such thing. Isaac Newton had a very charming niece who made a conquest of Montague. Fluxions and gravitation would have been of no use without a pretty niece." And curiously, it appears there may well have been some truth in this unlikely story.

Either way, Newton was not inclined to regard his job as a sinecure. He had other ideas. At the time the English currency was being heavily undermined by forgers and "clippers" (who snipped the edges off gold and silver coins). Once again, Newton had a legitimate target on which to vent his vast, suppressed rage. This time there was no danger (for him), and there was no stopping him. Within months Newton had become the terror of the London underworld, conducting a vindictive campaign against all the counterfeiters he could find. Over a hundred were flung into Newgate Jail, and Newton was responsible for a score of hangings at Tyburn. He insisted upon being present at all the trials.

Soon hardened criminals were literally trembling at the very mention of his name. Newton was out of control, touring the taverns (with an armed escort), conducting "interviews" with suspects and informers. During the course of these "interviews" he would give full vent to his fearsome rage—on hardened criminals and innocent alike. It was said that transcriptions of these interrogations, which presented only a formal version of the proceedings, read like *The Beggar's Opera*. Unfortunately Newton later destroycd these records— "of whch wee burnt boxfuls," according to the officer of the Mint who was his accomplice.

Newton's work soon began attracting attention beyond the confines of the underworld. A wealthy gentleman from Kensington called William Chaloner mounted a campaign against the Mint, accusing it of malpractice. He was known as an inventor, and suggested to Parliament that the Mint's coining machines should be replaced with an invention of his own. Ever averse to disclosing his methods of work, Newton refused point-blank to let Chaloner examine the Mint's coining machines. Whereupon Chaloner accused the Mint of making counterfeit coins, and being in league with the forgers.

This was a mistake. Newton had a lot to hide (heresy, etc.), and was paranoid about accusations of secret malpractice. Newton "investigated" Chaloner with merciless persistence, discovering from underworld informers that he had in fact made his fortune from counterfeiting coins. Although Chaloner had the protection of powerful friends, including well-placed MPs, Newton persisted with relentless passion. Chaloner was a ruthless man, betraying close colleagues, who were sent to the gallows, and issuing covert threats to Newton, in an attempt to elude justice. But Chaloner had called Newton a liar (which of course he was, every time he prayed in church). Newton could never rest while such a man lived. The result was inevitable. Chaloner was hung at Tyburn in 1699.

That same year, Newton was promoted to Master of the Mint, his salary increased to thirty-five hundred pounds. By now Newton had been forced to transfer his attentions to more serious matters. Already clipping had reduced the entire silver coinage to almost half its specified weight, and as a result English money was frequently refused on the Continent. This was playing havoc with trade, and the Treasury was on the point of collapse. If this happened, the Protestant monarchy was liable to follow suit, with a recall of the dreaded Catholic Stuarts.

In desperation, the government decided there was only one solution: The entire currency would have to be recoined. Newton applied himself to this herculean task with characteristic single-mindedness. At the Mint three hundred men and fifty horses (for turning the presses) were set to work, and six and a half million pounds of currency was recoined in three years. This was some achievement: Only half this amount had been produced in the previous thirty years.

In 1703 Hooke finally died, and Newton accepted the post of president of the Royal Society. Newton was incapable of magnanimity, and immediately ordered the burning of Hooke's portrait. He then set about revitalizing the society, which had declined into a mere gossip shop. He instituted weekly meetings, at each of which a new experiment was demonstrated. Unlike previous presidents, who seldom turned up, Newton was only to miss three meetings in the next twenty years.

Newton's presidency was marred by his characteristic behavior—only this time the victims were members of the Royal Society rather than the underworld. The most disgraceful episode concerned Flamsteed, the Astronomer Royal, whom Newton virtually hounded to his death.

Newton had crossed swords with Flamsteed while he was writing his *Principia*. In need of

observational figures to support his calculations of the lunar orbit, Newton had written to
Flamsteed at the Royal Observatory in Green-
wich. Flamsteed was a perfectionist, and had
already spent over a decade making observations for what was to prove the most accurate
and comprehensive map of the heavens yet
undertaken. Newton required accurate figures
immediately. Flamsteed was unwilling to let
the results of his long labors appear piece-
meal; and only released the figures reluctantly,
if at all. As a result, Flamsteed too eventually
suffered Hooke's fate—all the acknowledg-
ments to Flamsteed that Newton could find
were vindictively erased from the second edition of the *Principia*. But this was just for start-
ers.

When Newton became president of the
Royal Society, the Royal Observatory effectively fell within his domain. He immediately
ordered Flamsteed to publish all his findings
at once. Flamsteed protested, and for many
years fought a canny rear-guard action against
Newton's persistent bullying. But Newton was

not easily thwarted. Eventually he inveigled his friend Halley to seize Flamsteed's papers, his life's work. Halley was ordered to edit them himself for publication, and four hundred copies appeared. Flamsteed was outraged, as well he might have been, at this desecration of his precious work.

He managed to obtain an injunction against the Royal Society, which was publishing the work. But too late, copies had already been distributed. With the aid of friends, he managed to track down three hundred copies, which he personally burned. The Royal Society may have been the major institution of its kind in Europe, but its proceedings—especially those of its president—were sometimes far from scientific.

Now that Hooke was dead, Newton decided it was safe to publish more of his own work. (By then no one in England had the temerity to question Newton, let alone accuse him of plagiarism.) In 1704 Newton published his second masterpiece, the *Opticks*. This was in fact little more than a summary of the groundbreaking work on light that he had done thirty years previously. Appended to this work were two papers outlining his method of fluxions (calculus), which he had also discovered some thirty years earlier. Unfortunately, the German philosopher Leibnitz had published his own version of calculus twenty years earlier. The inevitable accusations of plagiarism soon mounted to a furor.

The facts, such as we know them, are as follows. Leibnitz had certainly seen some early papers of Newton's, but he had equally certainly worked out his own version of calculus independently. Indeed, his entire notation is different. (Leibnitz's notation—such as $\int$ for integration—is the one we use today, as is his name for his discovery: calculus.) Newton's unwillingness to face questioning over his work condemned his fluxions to history. Leibnitz's calculus was already being used by mathematicians on the Continent. Yet there's no doubt that Newton was the first to discover this method.

It soon became evident that the facts were

of little concern to either side. (This sadly unscientific approach to priority disputes among scientists was quickly to become established as a tradition that continues to flourish.) This time Newton had encountered an adversary even greater than Hooke—with an intellectual stature approaching his own, and an obduracy to match.

Leibnitz made the tactical error of accusing Newton of dishonesty, and once again Newton's anger knew no bounds. He became literally ill with rage. An acrimonious correspondence between members of the rival factions ensued, in the course of which the two greatest minds in Europe exhibited breathtaking unscrupulousness.

It was decided that the Royal Society should set up a committee to investigate the matter. Newton hijacked the committee's report, completely rewrote it in his own favor, and even anonymously reviewed it himself. He wrote numerous vituperative articles also in his own favor, which he browbeat other eminent scientists and mathematicians into publishing under their own names. Leibnitz did his own nasty best, and the controversy roared on until he died in 1716.

But Newton's rage was not so easily appeased. He continued to pursue Leibnitz beyond the grave. Visitors spoke of spontaneous tirades against the dead German philosopher, and practically every scientific paper Newton wrote from now on included a furious paragraph castigating his deceased adversary. Leibnitz was now unable to withdraw his accusation of dishonesty: So the stigma remained.

With some justification, psychiatrists have speculated upon whether there was some deeper dishonesty in Newton's nature that he felt compelled to conceal. Was there something beyond all the heresy, the pathological anxieties that drove him to work day and night, the inability to tolerate questioning or accusations of any kind? Perhaps. It's possible that he was terrified by intimations of suppressed homosexuality, or some other secret he couldn't face. Yet if so, his flight from this truth of his own nature drove him to discover far more profound truths about nature itself. His failure as a human being—both psychologically and in his behavior—would seem to have been inextricably bound up with his success in his work. (Though the unique quality of the latter remains beyond present explanation.)

The tendency has been to see Newton's later years as the waste of a great talent. True, he produced nothing—but what could he have produced? Such questions are usually otiose. Yet in Newton's case there was certainly unfinished business. Take his views on light, for instance. Against the emerging opinion, Newton clung to the ancient view that light consisted of a stream of particles. Yet he was willing to concede that certain evidence appeared to confirm the opposing wave theory. Had he continued to put his mind to this problem, it's not impossible that he would have arrived at something resembling the quantum theory of light, which views it both as particles and waves. (It was to be two hundred years before the Danish physicist Bohr came up with this theory, thus instigating twentieth-century physics.)

Newton's supreme genius certainly waned after he arrived in London at the age of forty-three, but he remained nonetheless a match for any mind in Europe. In 1696 Leibnitz had conceived of a problem, with the aid of his friend the Swiss mathematician Bernoulli, who had helped to develop Leibnitz' version of integral calculus. The problem was as follows: Two points are selected at random on a vertical plane. What curve does a heavy body follow when passing without friction under the force of gravity from the upper point to the lower point in the shortest time? This *Brachistochrone* (shortest time) problem was issued as a challenge to the leading minds of Europe. Newton received the details when he returned from work one afternoon after a hard day at the Mint. He was still averse to producing (or even discussing) his intellectual work in public, writing to a colleague: "I do not love . . . to be dunned & teezed by forreigners about Mathematical things." Despite this, he couldn't resist casting his eye over the problem after dinner, and by four in the morning he had solved it. The curve is a cycloid, the trace left by a point on the circumference of a circle as it rolls along a straight line. Next day Newton dispatched his solution anonymously; but Bernoulli knew at once who was responsible, making his celebrated remark: "I recognize the lion by his paw."

Twenty years later, when Newton was seventy-three, Leibnitz decided to have another dig at his old enemy—issuing a problem he pretended was a challenge "to the brotherhood of fine mathematical enquirers." (In modern parlance this asked: For any one-parameter family of curves, what are the orthogonal trajectories?) This question concealed a devious trap. Once again Newton received the problem when he returned from work at the Mint, and had solved it before he went to bed, circumventing the brilliant trap as an irrelevancy. Leibnitz refrained from comment; no further challenges were issued, and within a year Leibnitz was dead.

Newton's hair had turned gray when he was in his early thirties—"a Metamorphosis occasion'd by extremity of Concentration upon his Studies," according to a fellow don. But physically he was to remain in excellent health to the end of his days. He never needed glasses and he lost only one tooth. Contemporary sources claim that he lived frugally, even meanly. (French visitors to his home complained of "inedible fare, with paucity and poorness of the vintage in equal degree." But then, the French always made such remarks when visiting England.)

However, Newton's diet appears to have been frugal only by eighteenth-century standards. A typical weekly bill for household items includes a goose and a chicken, two turkeys and two rabbits; and when he died he owed no less than seven pounds and ten shillings to his brewer (itemized as fifteen barrels of beer). Such hardly speaks of saintly abstinence.

Newton remained Master of the Mint until the end of his life. He was also reelected annually to the presidency of the Royal Society (none dared oppose him). Even in his eighties he dutifully attended the weekly meetings, only occasionally dozing off during the proceedings. He also prepared later editions of his *Principia* and the *Opticks,* and continued avidly with his theological speculations, producing such works as *Observations Upon the Prophecies of Daniel and the Apocalypse of St. John* and *The Chronology of the Ancient Kingdoms Amended* (in which he calculated the exact date when the world had begun, according to his own interpretation of the biblical texts). He always found something to occupy him. And according to his psychological biographer, Frank E. Manuel: "Busyness, that false balsam of anxiety, took the form of obsessively copying when there was nothing else to do."

During his last years Newton was looked after by his niece—no longer a London beauty—and her husband, who conscientiously noted down the great man's stories and memories. Isaac Newton, the finest scientist who ever lived, finally died on March 20, 1727, at the age of eighty-four.

He was buried in pomp at Westminster Abbey, his pall supported by dukes, earls, and the Lord Chancellor, his funeral attracting vast crowds. Voltaire, who was visiting London at the time, marveled: "England honours a mathematician as other nations honour a king who has done well by his subjects."

Some Quotes

The rejected epitaph for Newton's tomb:
"Nature, and nature's laws lay hid in night.
God said, *Let Newton be!* and all was light."
                    —ALEXANDER POPE

Lines said to have been written on seeing Newton's statue in the chapel at Trinity College by moonlight:
". . . Newton, with his prism, and silent face:
The marble index of a mind for ever

Voyaging through strange seas of Thought,
alone.''
  —*The Prelude,* WILLIAM WORDSWORTH

Einstein's introduction to a new edition of the
Opticks:
''Fortunate Newton, happy childhood of science! . . . Nature to him was an open book, whose letters he could read without effort. The conceptions which he used to reduce the material of experience to order seemed to flow spontaneously from experience itself, from the beautiful experiments which he ranged in order like playthings and describes with an affectionate wealth of detail . . . Newton's age has long since been passed through the sieve of oblivion, the doubtful striving and suffering of his generation had vanished from our ken; the works of some great thinkers and artists have remained, to delight and ennoble us and those who come after us. Newton's discoveries have passed into the stock of accepted knowledge.''

Newton's definition of force from the *Principia*:

"An impressed force is an action exerted upon a body, in order to change its state, either of rest, or of uniform motion in a right line.

This force consists in the action only, and remains no longer in the body when the action is over. For a body maintains every new state it acquires, by its inertia only. But impressed forces are of different origins, as from percussion, from pressure, from centripetal force."

Leibnitz' opinion of Sir Isaac Newton before they fell out:

"Leibnitz said that taking Mathematicks from the beginning of the world to the time of Sir I. What he had done was much the better half—and added that he had consulted all the learned in Europe upon some difficult point without having any satisfaction and that when he wrote to Sir I. he sent him answer by the first post to

do so and so and then he would find it
out."

—In reply to the Queen of Prussia:
Keynes MS.

From Newton's account of planetary motion:
"Centripetal forces are directed to the individual cen-
tres of the planets.

That there are centripetal forces actually
directed to the bodies of the sun, of the earth,
and other planets, I thus infer.

The moon revolves about our earth, and
by radii drawn to its centre describes areas
nearly proportional to the times in which
they are described, as is evident from its velocity compared with its apparent diameter;
for its motion is slower when its diameter is
less (and therefore its distance greater), and
its motion is swifter when its diameter is
greater.

The revolutions of the satellites of Jupiter
about that planet are more regular; for they
describe circles concentric with Jupiter by uni-

form motions, as exactly as our senses can perceive.

And so the satellites of Saturn are revolved about this planet with motions nearly circular and uniform, scarcely disturbed by any eccentricity hitherto observed."

Chronology of Newton's Life

1642   Born in hamlet of Woolsthorpe,
       Lincolnshire

1644   His mother remarries and moves to
       nearby village, leaving Isaac to be
       brought up by his grandmother

1653   Mother returns home after death of
       her second husband

1659   Mother calls Isaac back from school
       in Grantham to look after the farm

1661   Goes to Trinity College, Cambridge,
       as a sizar

1665   Receives B.A. from Cambridge and
       flees back home to Woolsthorpe to
       avoid the Plague

1665–6	Newton's *annus mirabilis,* during which he receives the inspiration for his law of gravity
1667	Returns to Cambridge, elected Fellow of Trinity College
1669	Becomes Lucasian Professor of Mathematics at Cambridge
1672	Elected a Fellow of the Royal Society
1678	Suffers first nervous breakdown after controversy with Hooke
1687	Publishes *Principia Mathematica*
1693	Suffers mental collapse after break with Fatio
1696	Moves to London and becomes Warden of the Mint
1699	Promoted to Master of the Mint
1703	Accepts presidency of the Royal Society on the death of Hooke
1704	Publishes *Opticks*
1727	Dies at the age of eighty-four

Chronology of Era

1642   Death of Galileo
        Outbreak of civil war in England
        Dutch explorer Tasman maps coast
        of Van Diemen's Land (now
        Tasmania)

1648   End of Thirty Years' War, leaving
        large tracts of Germany devastated

1649   Execution of Charles I and
        proclamation of the
        Commonwealth, the first successful
        revolution and establishment of a
        republic in a major European
        country

1650   Death of Descartes

1660   End of the Commonwealth and
        Restoration of Charles II to the
        throne

1664–5 Plague spreads through England

1665   Death of great French
        mathematician Fermat

1669   English government draws up
        constitution for the colony of the
        Carolinas in America

1688   Glorious Revolution: James II flees
       the country and the Protestant
       monarchs William and Mary ascend
       to the throne

1690   Locke publishes *An Essay Concerning
       Human Understanding,* instigating
       philosophy of empiricism

1706   Act of Union between English and
       Scottish Parliaments establishes
       Great Britain

1715   Louis XIV, the "Sun King" dies at
       Versailles

1716   Death of German mathematician-
       philosopher Leibnitz

Suggestions for Further Reading

Christianson, Gale E. *Isaac Newton and the Scientific Revolution* (Oxford University Press, 1996) Far from reverential, and scientifically detailed.

Cohen, Bernard (ed.). *Newton* (N. W. Norton, 1996) Texts, backgrounds, and commentaries.

Hall, A. Rupert. *Isaac Newton: Adventurer in Thought* (Cambridge, 1996) Newton's contributions to history, theology, chemistry, and philosophy.

Manuel, Frank E. *A Portrait of Isaac Newton*

(DaCapo Press, 1990) The best psychological study.

Peterson, Ivars. *Newton's Clock* (W. H. Freeman Co., 1995) Newton and beyond: Chaos in the universe.

Westfall, Richard S. *Never at Rest* (Cambridge, 1996) The definitive modern biography.

Westfall, Richard. *The Life of Isaac Newton* (Cambridge, 1994) A condensed edition of *Never at Rest.*

the key to understanding the cosmos, despite the devastating effects of motor neuron disease. Brilliantly simplifying the most complex ideas, *Hawking and Black Holes* will help you grasp the universe in ways you never thought possible.